WILDLIFE
of Western Canada

WILDLIFE

of Western Canada

A Photographic Portrait
by Dennis and Esther Schmidt

Altitude Publishing
The Canadian Rockies/Vancouver

PUBLICATION INFORMATION

An Altitude NatureBook

Copyright © 1995 Altitude Publishing Canada Ltd.

Canadian Cataloguing in Publication Data

Schmidt, Dennis, 1921-
Wildlife of Western Canada
ISBN 1-55153-084-8 (bound) -- ISBN 1-55153-085-6 (pbk.)
1. Zoology--Canada, Western--Pictorial works. 2. Wildlife photography--Canada, Western.
I. Schmidt, Esther, 1922- II. Title.
QL221.W47S37 1995   591.9712   C95-910304-X

Concept/art direction	Stephen Hutchings
Design	Stephen Hutchings
Text editing	Nancy Flight
Author interviews	Alison Barr
Electronic page layout	Sandra Davis and Alison Barr
Altitude NatureBook Series Editor	David Spalding
Financial management	Laurie Smith

Printed in Canada
Printed and bound in Western Canada by Friesen Printers, Altona, Manitoba.

Altitude GreenTree Program
Altitude Publishing will plant in Canada twice as many trees as were used
in the manufacturing of this book.

Front cover Grizzly
Pages 2–3 Snow geese
Pages 4–5 Elk
Page 6 Grizzly

Altitude Publishing Canada Ltd.
The Canadian Rockies
P.O. Box 1410, Canmore Alberta T0L 0M0

Dennis Schmidt photographs a bighorn ram

Contents

Wildlife in Western Canada

Since Europeans first began exploring North America, our attitudes towards wildlife and the natural world have undergone dramatic—and necessary— changes. A few centuries ago, the western reaches of the continent seemed vast, limitless, bountiful, full of unending natural wealth. The idea of "the West" was synonymous with "frontier"—a place where personal dreams could be realized through hard work, where right and wrong were directly asserted by one's own personal strength and power. In this country, the equation was simple: "Western Canada" equalled "huge skies, enormous mountains, raging rivers, thundering oceans"—and wildlife in fantastic numbers and varieties, from buffalo herds to pelican flocks to endless schools of salmon. For the early European immigrants, the sheer quantity of the wildlife was overwhelming, almost unimaginable, limitless. In response to this abundance, they concluded that there was huge economic gain to be had from the animals. Because the number of animals appeared to be so vast, hunting was pursued—whether for food, skins, bones, or even the pleasure of the kill—as if it would make no difference to the overall size of the animal populations.

We now know that such actions and attitudes have had a devastating effect on the wildlife of western North America. At the end of the 19th century, in the space of only 20 years, the buffalo were reduced from over 20 million (some reports estimate the number to have been closer to 50 million) to under 1,000 animals. The same story is true for grizzlies, for wolves, for bald eagles, for many of the species hunted and slaughtered into virtual extinction. In North America over 500 species of all kinds have suffered extinction in the past 500 years.

A resurgence is at hand, however. Although this resurgence is not spectacular, there is a substantial increase in real numbers of many species of North American wildlife, such as buffalo, bald eagles, and wolves, which had previously been brought close to extinction. The reason for this turnaround is the change in attitude of the human population. Whereas human behaviour towards the animal world in the past was driven almost completely by the desire for economic gain and personal consumption, our current behaviour is characterized by such attitudes as "enjoyment" and "understanding," attitudes that are now seen as valid objectives in their own right, not merely side benefits of the exploitation of wildlife.

This shift in attitude towards wildlife is due not only to the maturing and deepening of our understanding of the natural world but also to the development of the camera and the increasing interest in and demand for visual images of the untamed world. Photography has made it possible to shift society's view of wildlife from one rooted in consumptive economics to one that is essentially cultural and recreational. In other words, with respect to animals, the camera represents a triumph of understanding over economics.

Opposite Dennis Schmidt photographs grizzly tracks in the Canadian Rockies

This is precisely the strength of the work of Dennis and Esther Schmidt. Their vision of wildlife in Western Canada is a result of their attempts to understand the animals within the context of the animals' own natural environment. The Schmidts' photographs are the result of painstaking efforts to establish a relationship with their subjects within an atmosphere of trust and acceptance. This book is a direct result of these efforts; it is an intimate, "behind-the-scenes" portrait of the animals as they move about their world.

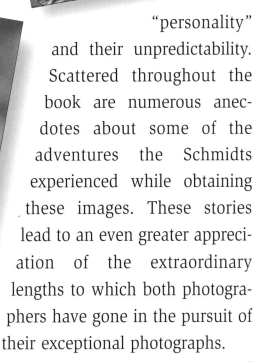

Dennis and Esther's work with their animal subjects was not unidirectional but interactive, and the animals frequently surprised the Schmidts with their

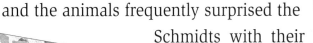

"personality" and their unpredictability. Scattered throughout the book are numerous anecdotes about some of the adventures the Schmidts experienced while obtaining these images. These stories lead to an even greater appreciation of the extraordinary lengths to which both photographers have gone in the pursuit of their exceptional photographs.

Considering our not-so-proud history with the animals that share this part of the world with us, it is to the Schmidts' credit that they have created and assembled such a fine collection of photographs. It is also to their credit that they have developed and nurtured, over a period of years, a sensitive and understanding relationship with the wildlife of Western Canada.

Stephen Hutchings

Top Esther Schmidt
Bottom Dennis Schmidt

Relating to Wildlife

Dennis and Esther Schmidt have developed a relationship with wildlife based on respect and admiration. Right from the start, they try to understand the animal they intend to photograph by spending time in its territory, getting to know the land, the waterways, where the animals can be found, the trails they use, where they feed, and so on. Then, when the conditions are right for photography, they approach the animals with a differential attitude so that the animals will remain comfortable and unthreatened by the "intruders." In all cases they regard the territory as belonging to the animals; it is the Schmidts who must show obeisance.

"To be honest, we feel far less threatened by animals than humans," says Dennis. Just as most people would not like a stranger to burst into their living room, the Schmidts do not barge onto an animal's home ground.

Some of the techniques they use to help their subjects feel less threatened include showering with unscented soap and using unscented deodorant, moving slowly and carefully in the animals' territory, and using submissive body language in the presence of the animals. In spite of their precautions, if the animals do object, then the Schmidts respect their wishes and withdraw from the area. More often than not, the animals show little concern for their presence. Understanding the animals' intelligence and social

Top Esther Schmidt with a rattlesnake
Bottom Dennis Schmidt and a ground squirrel

13

structure has also helped the couple be accepted by the various kinds of wildlife in which they are interested. Learning from the animals themselves about their habits, preferences, and territory has been part of the process of photographing wildlife. The more the Schmidts learned about a certain species, the easier it was to fit into its world and to take photographs.

Occasionally the signals have become mixed and the Schmidts have been chased by an upset animal. (While one photographer scrambles for cover, however, the other is apt to be photographing all the action!) The Schmidts are quick to point out that in these cases the blame for any aggressive behaviour is entirely theirs.

"Take the picture first; then worry about identification and safety later," is Dennis and Esther Schmidt's motto for photographing wildlife. Both photographers have been asked repeatedly why they don't carry a weapon, such as a rifle, for protection, even if they don't intend to use it. Their

Top Esther Schmidt with a gray jay
Middle Dennis Schmidt at a beaver dam
Bottom Esther Schmidt photographing a tide pool
Opposite Dennis Schmidt approaches a mountain goat

explanation is that animals can sense if a person has ill intentions towards them. A weapon would introduce an aura of fear or apprehension into the situation and produce the opposite result from what they intend. "Animals instantly sense fear; when you are afraid, your skin gives off an odour and the animals know it right away," says Esther.

In situations where the Schmidts, for one reason or another, were unable to photograph—perhaps because the animals had been scared away or because the lighting conditions were bad—they would sit down, relax, and enjoy the scenery. Once, just as dusk was falling after a hard day's work and they were relaxing at the edge of a meadow, they watched a coyote hunt for gophers. After the coyote had made a number of unsuccessful attempts to catch a gopher, a badger joined the hunt—not as a competitor, it turned out, but as an ally. The coyote would dig into one of the gophers' holes, forcing the gophers to retreat underground and pop up at another exit—only to be grabbed by the waiting badger! The process would then be reversed, with the badger digging and the coyote catching the bewildered gophers.

After a lifetime love affair with photographing nature, Dennis sums up the Schmidts' experience by saying: "We have been shooting part time since we were kids and full time for twenty years. It is the longest job we've ever had, and we enjoy it more now than ever."

Opposite Canada goose
Top Esther Schmidt with bighorn rams
Bottom Esther Schmidt photographs a Canada goose

17

Pages 18–19 Sandhill cranes
Opposite Mountain bluebird
Top Buffalo **Bottom** Beaver

21

Pronghorn Antelope

We were on a large, rolling grassland in southern Saskatchewan when we saw a large object ahead of us. We thought it was probably a rock, which was good, because we knew that rocks were frequently used by animals either as a home base or for protection. As we got closer we saw that sure enough, lying beside the rock was a baby pronghorn antelope. It hadn't been abandoned—the mother had left it hiding there while she went to feed—we could clearly see her staring at us from the top of a knoll. We got some terrific shots of this little creature. He stayed right there and didn't move.

Dennis Schmidt

Opposite, top Mountain goats
Opposite, bottom Badger
Above Pronghorn antelope

Opposite Turkey vulture
Above Wild turkey

Opposite Harbour seal
Top Starfish feeding on mussels
Bottom Great blue heron

Black Bears

We were out hiking and came across a sow black bear and two cubs that were looking for food near a beaver dam. Eventually, the sow, followed by her cubs, gave up the search, crossed through the woods, and headed into the trees. All three of them proceeded to climb up a poplar tree to feed on the buds. It was amazing to see these three bears near the top of this small tree, nibbling on the succulent leaf buds. I remember it as a very peaceful time in which the bears knew we were there but nonetheless didn't feel the least bit threatened.

Dennis Schmidt

Page 28 Black bear
Page 29 Moose
Opposite Black bear
Above Grizzlies

Opposite Cedar Waxwings
Top Common loon
Bottom Cinnamon teal

Porcupine

One autumn a porcupine was out feeding on dandelion leaves in a field with no trees around. We were working on either side of him, when he suddenly headed for the nearest willow bush. Since the bush was not big enough to climb and the porcupine couldn't escape our attentions, Dennis decided to try to take some portrait photographs with a 50 mm lens. Dennis got down on his hands and knees and crawled to the willow bush beside this little critter. Luckily, the porcupine was completely docile and Dennis was in no danger of getting lashed. While he was photographing this animal, talking to him all the time, his left arm touched the porcupine's body—obviously he accepted Dennis. After the pictures were taken, Dennis put down his camera and stroked the nose of this little porcupine, who was now totally relaxed and seemed to be enjoying all the attention.

Esther Schmidt

Pages 34–35 Gray wolves
Opposite, top Skunk
Opposite, bottom Raccoon
Above Porcupine

Red Fox

To be accepted by a wild animal is one of life's greatest moments for me. It was autumn and we had found a gorgeous red fox. We shot off several roles of film in just a few days. Then one day he truly accepted us. He walked to within 10 feet of us and quietly stopped. I tried to get a picture, but there were several blades of grass in front of his face. "What should I do?" I wondered. I decided to get down on my belly and crawl towards him in order to remove the grass, which was only a couple of inches from his eyes and his gleaming white teeth. He let me reach out and snip the blades of grass without so much as a snarl or a flinch on his part. That night, after we went back to our camper, which was about a mile away, there was suddenly a strange noise at the door. We opened it, and to our surprise, there was our little fox, obviously curious about us because he had followed us home.

Esther Schmidt

Page 38 Osprey
Page 39 Bald eagle
Opposite, top Great horned owl
Opposite, bottom Cougar
Above Red fox

Opposite Green heron
Top White pelicans
Bottom Salmon

Opposite Frog
Top Pacific rattlesnake swallowing a deer mouse
Bottom Painted turtle with damselfly

Bighorn Sheep

Rocky Mountain bighorn sheep are usually quite amiable and easy to work with. By taking it slow and being patient, we can usually get within range to get good pictures. There are always exceptions, however. Once, when we were photographing a herd of the animals, mostly rams, one ram took a dislike to Esther. No matter where she hid—behind rocks or trees—he went after her. Everywhere she went, he followed, trying to attack. Finally, having had enough of this, we both decided to leave. Even in confronting a situation such as this, however, Esther insists she has no fear of animals.

Dennis Schmidt

Pages 46–47 Osprey in their nest
Above Bighorn ram
Opposite Bighorn rams in collision

Top Pika
Below Yellow-pine chipmunk
Opposite Coyote

Black Bear

In northern Saskatchewan, we came upon a pretty marsh while hiking along a game trail that paralleled a creek. We sat very quietly on the sunny hillside above the marsh, relaxing and talking and passing the time. Suddenly, a very large black bear jumped out of the bulrushes. He had obviously caught our scent, but he was confused because he couldn't see us. He paused, looked around, and then headed back into the forest. I warned Esther to watch carefully—the bear seemed to have something up his sleeve. After about 15 minutes we saw him cross the marsh about a quarter of a mile downstream. He made his way towards us on our side of the marsh but disappeared before he reached us. Apprehension started to build as we waited and waited—sensing that he was watching us—but we couldn't see him. Finally, after quite some time, we slowly and cautiously started to walk back to the camper. We hadn't gone more than 100 feet when we saw the bear peeking out at us from behind a small pine tree. He stalked us all the way back to our camper.

Dennis Schmidt

Opposite Grizzly bear with cubs
Above Black bear cubs

Moose

One winter day we discovered moose tracks and followed them for some distance, eventually coming upon a cow moose with her calf, lying in the snow ruminating. It was deep snow, knee deep, which is not deep for a moose, but it was for us. There was a beautiful backdrop and I wanted to keep the animals as large as possible in the frame of my camera, so I decided to use a shorter lens and get as close as I could. But I went one step too near. Both animals were on their feet in flash, and they weren't heading in the other direction. The cow came directly towards me! I know she didn't have any ill intentions towards me—other than to scare me into leaving—but as I stepped aside I could feel the rush of air as she and her calf whistled by.

Dennis Schmidt

Opposite Moose feeding in a lake in the Canadian Rockies
Above Moose cow

Coyote

We discovered the territory of a beautiful coyote and decided to stake it out over a period of time, waiting for him to accept us without a look of fear on his face. Each day he came closer and closer; a week passed, and then finally he lay down just 20 feet from me—and I promptly ran out of film! Instead of going for more film, I decided to stay where I was and see what would happen. I knelt on the ground with my hands folded in front of me. Several minutes went by, during which the coyote and I watched each other intently. Slowly he came closer and closer until he was right in front of me. His head reached towards me, and he opened his mouth, took my hands in his mouth, held them gently for several seconds, then released them and retreated to within 15 feet. He then lay down and we quietly enjoyed each other's company for about 10 minutes. Finally, he rose and sauntered into the solitude of the surrounding woods. It was an incredible experience.

Esther Schmidt

Opposite, top Burrowing owls
Opposite, bottom Pronghorn antelope
Above Coyote chasing a deer mouse

Opposite Northern sea lions
Above Harbour seals

Elk

One winter day I saw a beautiful scene with the sun and the mountaintops peeking through the clouds and half a dozen cow elk in a field. The snow wasn't too deep and the ice on the river was strong enough to carry me, so I headed across the river into the pasture with these elk. The female nearest me was a large, magnificent beast. I had a 35 mm lens on my camera, so I had to move closer and closer to get this beast large enough in the frame. When I got within 75 feet, she took exception to me and charged. There was no point in running and nothing to hide behind in the open pasture except one tiny little 4-foot-tall fir tree. I quickly stepped sideways behind this tree, and for some reason—for it was no real obstacle to her—she stopped in her tracks, ears back, breath snorting out of her nostrils. I stood behind the tree frozen with fear. She slowly calmed down and walked away. I was able to take a photograph, and every time I look at the resulting picture, I still get a chill. A shorter lens has a much different perspective than a telephoto lens, and that is why I tried to get closer. By reading the animal's language, we have a good idea if we have gone too far. Once in a while we might miscalculate, but we have never intentionally endangered ourselves or the animals.

Dennis Schmidt

Previous page left Great horned owl
Previous page right Goshawk
Opposite Bull elk bugling
Above Bull elk

Pages 64–65 Bull moose in the snow
Opposite Marten
Top Bobcat kitten
Bottom Bobcat

Opposite Brown pelican
Top Red-necked grebe
Bottom Black tern with babies
Pages 70–71 Red-necked grebes at the nest

Wolf

Some bighorn ewes and their lambs were grazing on a hillside. I hiked up to see what photographs I could get, and I worked with the animals for some time. At one point, they all turned to look at a pack of coyotes crossing the ice below. I was even able to take a picture of the sheep with the coyotes in the background. When it was time to leave, I started back down, only to discover wolf tracks inside the footsteps that I had left on the way up. So I turned around to follow the new tracks, hoping to get some shots of the wolf. As it turned out, I followed the tracks to a spot about 100 feet above where I had been when I was photographing the sheep. Obviously, the wolf had been sitting here, silently watching the whole process, and when I had started back down the hill, he headed off as well. Finding that wolf's footprints in mine was enough to make my day. It isn't just a great photograph that makes a trip memorable; it can often be a small event, such as this one, that leaves a lasting impression.

Dennis Schmidt

Page 72 Red fox
Page 73 Red-tail hawk
Opposite, top Mountain goats
Opposite, bottom Dall sheep
Above Wolf

Deer Mice

It was early winter and we were searching for an interesting subject. We looked outside and saw a coyote jumping through the snow, hunting mice, and this gave us an idea: if we trapped a live mouse, we could see if we could entice the coyote into some more action the next morning. We were able to trap some mice all right, but it snowed almost 2 feet that night and we couldn't find it in our hearts to take the mice outside and let them go, because they would never find their way home. So Dennis built a place for them in our basement with all of the amenities of home—even a little exercise wheel. This presented a great opportunity to photograph these charming little rascals throughout the winter. We let them go in the spring.

Esther Schmidt

Opposite Deer mice
Above Great horned owl

Bull Elk

I **was charged by a large bull elk** one day, an event that was entirely my own fault. I had been following this magnificent elk for about a mile to get the "perfect photograph." A close-up of it bugling would be great. I got closer...he got closer...I got closer...he got closer...until there was only about 8 feet between us. Suddenly his ears went back, and he lowered his antlers and charged! Did my adrenaline ever flow then! I held my ground and at first he gave way, but then he charged again. I stepped behind a small tree—a bush, really—and he gave me a disdainful look and wandered off. I did not follow after him this time.

Esther Schmidt

Opposite Bull elk
Above Bald eagles

Cow Elk

While driving along a back road through the Alberta countryside, we rounded a bend in the road and came upon a herd of cow elk with their young. We stopped our camper truck and rolled down the window, and I positioned my camera on the window mount that we carry with us for this kind of situation. Gradually the elk lost their fear and advanced closer and closer until they surrounded the camper. Suddenly one female came up to the vehicle and stuck her head into the window, past the camera on its mount, and gave me a wonderful wet kiss right on the side of my face.

Esther Schmidt

Opposite, top Golden-mantled ground squirrel
Opposite, bottom Red squirrel
Above Elk cow with calf

Above Northern flicker
Opposite Yellow-bellied sapsucker

Porcupine

It was late in the evening during October and there was a fair amount of snow on the ground when I spied a mother porcupine and her son digging out dandelion leaves through the snow. I got down on my belly and dragged myself along until I was 12 feet behind them. That's when I discovered they were talking to each other. I thought that this was very interesting and wondered if I could talk like a porcupine. I decided to give it a try and made some sounds imitating the porcupine as best as I could. I must have said the wrong thing, however, because the mother got mad—at her son, not at me! She gave him a terrific scolding. Finally she saw me, and then her little beady eyes swelled to great proportions and in no time both she and her son were gone.

Dennis Schmidt

Opposite Porcupine
Above Black-tailed prairie dogs
Pages 86 and 87 Grizzly bears

Beavers

Late one summer we watched some beavers that decided to build a new lodge on a lake. They seemed to realize that winter was closing in and that there was little time left to build the lodge as well as put enough food away for the winter. Instead of just working at night, which was their normal behaviour, they also started working during the afternoons. We were often there taking photographs, so these beavers became used to us. They would come close enough to smell our shoes before starting to fell trees for their lodge and for the winter's food supplies. Pretty soon it got to be a ritual: each day they would sniff for us, smell our shoes, and start to work.

Esther Schmidt

Pages 88–89 White tailed deer
Opposite and above Beaver

Above Orca (Pacific killer whale)
Page 94 Snowy owl
Page 95 Great horned owl

Page 96–97 Woodland caribou
Opposite White-breasted nuthatch
Top Barn swallow with nestlings Bottom Common nighthawk

99

Gray Wolf

While hiking along a small creek in northern Saskatchewan we found wolf tracks on a sandbar and, a little farther up the creek, the remains of a beaver—fresh remains, including the beaver's skeleton and feet. So I took a few pictures of the feet to show the grooming claw on the back feet that the beaver uses to clean its coat and restore the oil that makes its coat water-repellent. After that we continued our hike along the creek bed and came to a cliff about 10 or 15 feet high. We could see more tracks in the sand at the foot of the cliff. I wondered what would happen if I gave a wolf call, so I called—and there was an immediate response. There, on the top of this little bank, appeared a beautiful gray wolf. It totally ignored us as it looked for the source of the call; it didn't acknowledge us at all. As this noble beast looked off into the distance we were each able to take a couple of photographs before it pulled back and disappeared.

Dennis Schmidt

Opposite Bighorn lambs
Above Gray wolf pup

Black Bear

One day we came across a particularly beautiful bear that was obviously in a bad mood and wanted to be left alone. Because he was so attractive, we wanted to take some pictures. Within seconds the bear let us know that he wasn't in the mood for that sort of thing. He put his ears back and charged Dennis three times. Dennis held his ground, making as much noise as possible by snapping the legs of his tripod together and shouting at the top of his voice. The bear backed off and I was the one who got the pictures. Another day we startled a black bear while he was feeding in a berry patch. He was just as surprised as we were and burst into a fast run. It was the only time we have seen the underpads of all four bear paws at one time—suspended well above ground as he dashed for the trees.

Esther Schmidt

Opposite Black bear
Above Grizzly bears playing

Top Fisher
Bottom Hoary marmot
Opposite Wolverine

Pileated Woodpecker

I **had a terrific experience** one time when I was shooting a video of pileated woodpeckers. These birds habitually stake out a territory and stay there all year long. They also make long roosting holes in which to spend the night. In this instance, the male was digging himself a new nesting hole and I was shooting these pictures on video. Suddenly the woodpecker jumped away from the hole and was immediately grabbed by a hawk that had swooped down into the scene. The woodpecker flew in my direction with the hawk clinging to his back. The sharp-shinned hawk landed just behind me with a look of sheer terror in his eyes—but he wasn't going to let go. I felt I could almost read his mind: "I have dinner here and I'm not going to give it up, even if there is a human standing right in front of me!" Just then the female woodpecker flew in and landed on the hawk's back. What a commotion there was! All three of them flapped and fluttered to about 20 feet away from me. The female woodpecker simply wouldn't let go, and she eventually forced the hawk to give up. Unfortunately, the video camera was trained on the woodpecker hole, so I didn't get the shots of the three of them together. But I still vividly remember the chilling, sometimes bloodcurdling sounds of the action.

Dennis Schmidt

Opposite Pileated woodpeckers
Above Hummingbird on the nest

Opposite California quail
Top White-tailed ptarmigan, summer plumage
Bottom White-tailed ptarmigan, autumn plumage
Pages 110 and 111 Mountain lions

Dennis and Esther Schmidt

PHOTO CREDITS

Esther Schmidt
9, 10, 13 (bottom), 14, 15 (middle), 16, 20, 21 (top),
23, 24, 27 (top), 30, 32, 33 (top and bottom), 36 (top), 43 (top),
49 (middle & bottom), 55, 56 (top and bottom), 60, 66,
67 (top), 69 (bottom), 72, 74 (top and bottom),
77, 78, 80 (top), 81, 83, 84, 85,
94, 96/97, 98, 99 (bottom),101, 104 (bottom)

Dennis Schmidt
2/3, 4/5, 6, 12 (top), 13 (top), 15 (top and bottom), 17 (top and bottom),
18/19, 21 (bottom), 22 (top and bottom) 25, 26, 27 (bottom),
28, 29, 31, 34/35, 36 (bottom), 37, 38, 39, 40 (top and bottom),
41, 42, 43 (bottom), 44, 45 (top and bottom), 46/47, 48, 49 (top)
50 (top and bottom), 51, 52, 53, 54, 57, 58,
59 (top and bottom), 61, 62, 63, 64/65, 67 (bottom), 68,
69 (top) 70/71, 73, 75, 76, 79, 80 (bottom), 82,
86, 87, 88/89, 90, 91, 92/93, 95, 99 (top), 100, 102, 103,
104 (top), 105, 106, 107, 108, 109 (top and bottom), 110/111, 112

Brian Varty
12 (bottom)